T0196564

THE
FORBIDDEN
INSTRUCTION BOOK

Neil Riding

Order this book online at www.trafford.com
or email orders@trafford.com

Most Trafford titles are also available at major online book retailers.

Printed in the United States of America.

ISBN: 978-1-4669-4506-7 (sc)
ISBN: 978-1-4669-4502-9 (e)

Trafford rev. 06/27/2012

 www.trafford.com

North America & international
toll-free: 1 888 232 4444 (USA & Canada)
phone: 250 383 6864 ♦ fax: 812 355 4082

CONTENTS

PREFACE

The momentum for this book comes from two video clips; which are about saving students from wasted-learning; in particular during school hours. The two "documentary-style" clips were produced by 'Richard Dawkins'; a well-known English scientist.

["The god delusion" explores the unproven beliefs that are treated as factual by many religions, and the extremes to which some followers have taken them. Dawkins argues that "the process of non-thinking called faith" is not a way of understanding the world, but instead stands in fundamental opposition to modern science and the scientific method, and is divisive and dangerous. "The Root of All Evil" is a television documentary, written and presented by Richard Dawkins, in which he argues that the world would be better off without religion. The documentary was first broadcast in January 2006, in the form of two 45-minute episodes on Channel 4 (in Britain)].

Richard Dawkins' interest in these matters took him into the dark places of the world and led him to ask some difficult questions. He put himself 'in the picture', and he seems to get the point across. And yet so many have watched; and then just walked away, and forgot the message.

I am inviting you to watch—read the text—and consider the outcome. The overall title for his work is fair; although it's more likely that others were involved in the naming process. I hope that, by the time you have read through *this* material, you will like my suggested title, which is—'**A Dummies Guide to False Religion**'. His work is basically made up of two parts . . .

Part 1. "The God Delusion"
Part 2. "The Virus of Faith"

The clips are available to watch on-line. Some of his thoughts are repeated in this book, (to make sections easier to follow).

In these two brief videos—Richard Dawkins gives an adequate account of how "they" appear to the outside world. My aim is to re-visit his material, and take the 'Instruction-Book' along for the ride. All the while; trying to show—that the religions of this world are really self-made; rather than 'real religions' following a 'real book.' One of the most obvious threats to the credibility of "religion," is the fact that there are hundreds of churches—as well as all their (different) offshoots.

If the various churches were in fact following a similar Instruction Book, they would each have a similar internal structure. For example they would each have the same position; (title); for the various levels of competence within their ranks. The present system has all sorts of (titles); for their workers and all sorts of ideas about how things (must) be done. Also; to suggest that the religions might have 'made up' the Book (for their own benefit), would be utterly false, because they couldn't possibly have done a worse job. It simply wouldn't make sense, for any religion to 'invent' a book that completely opposes their practices; and their 'method-of-operation.'

If you already dismiss the book, you can see that no good religion could ever come from such a thing, but you may be interested in the difference between the world-view and the *actual* text.

If you still find yourself believing the religions simply 'made up' a fancy bunch of books, to set themselves up for their own "belief-system," then you should easily change that view when you realize the vast **disconnect** between the book and the end product (i.e. false-religion).

Regardless of where you now stand on religion; you will soon come to realize; that the religions are a *man-made* occurrence, and therefore cannot be built from their "**claimed**" Instruction Book. The religions are simply using the book as a leaning post, perhaps using it as an excuse for their actions.

In the future, the question will arise (about removing man-made religions from our education system). Please consider some of the points that are made throughout; when you make your decision on such an important matter.

There will be parts; where the reader might see that some things are difficult to cover. My first aim is to stick to the important topic, which is to get the message through to everyone. The stuff being taught in the name of religion—really is a waste of time. The students are learning things that have to be 'unlearned' at a later date. For example, students would need to unlearn the false; in order to take up a logical study of science. Students would need to unlearn the false; in order to just live a "quite and peaceful life," (after their education is complete). And worst of all, students would need to 'unlearn' the false in order to have any chance of understanding the "Forbidden Instruction Book." Any misquotes about the Instruction Book, might go unanswered; as present day thoughts could be the result of long held views; (which is the very thing that needs to be cleared up). Presentations; which are pro-evolution are generally ignored. Another thought that comes to mind is; "why would I use the term—*forbidden book*"? The answer comes from my own experience during my school years. Someone had challenged me to "go look it up." So I toddled off with a scrap of paper, with a chapter name and some directions. I dusted off an old family Book and began to flick through, looking for an exact chapter and verse, (to see what it really said). Then I heard others heading towards the room, I quickly turned the book over and sat a magazine on top. It seems I didn't want anyone to see me reading *that book*. When I thought about it later; it seemed quite strange. "Why was that one book so different from any other?" If it were another book, say a history book, or a book of genealogy or even a novel, I would not have felt any stigma whatsoever. I realized that people like me were actually forbidding themselves from reading it. I began to wonder what kind of silent force was at work (inside my head) to make me feel that way. And why was it only *that book.*

The next thing to consider is how to work out your own rating for "science verses instruction;" Using a blank sheet of paper, write across the top . . . "The Table of Truth." Then, divide the page into three columns.

1. Science. 2. Instruction 3. Religion.

At various points you can place a score in the columns,

1 — 1 — 0 and so on.

You can apply critical-thinking when writing your "score." For instance; if the scientist is re-stating something of a religious nature and the 'Book' disagrees with the religious statement, that could be scored as follows, science one — instruction one — religion zero.

If a religious fundamentalist makes a statement or claim and the Book has a near-opposite message, then that would be scored as, science one — instruction one — religion zero.

Or if the scientist makes a statement; that is his own opinion; (or of a "scientific" viewpoint), *and* the Book agrees with the scientist, then, Science one — instruction one — religion zero, and so on.

Keep the table in mind as you **check** information.

You must not make a score in your table based on my opinions or notes. You should only count your own opinion based on what you "see."

In the end, consider how the table looks. If you have two columns with a score of one and the final column, all zeros, then it may be time to admit that there is a gigantic divide between instruction and religion; (and perhaps your previous 'world-view' was a little rusty).

Someday, the opportunity will arise for you to vote in a referendum, containing questions about "religious" studies within the education system, please drag out your "Table-of-Truth" as a reminder (when the time comes). Once enough people realize the sheer waste involved; in learning and then unlearning the false things in life, they may work together to bring about important (and much needed) changes to our value system, and free-up student's time for learning something worthwhile. Many of the quotes throughout are from Richard Dawkins work, (which is available to watch on-line), with the titles "The god Delusion' and 'The Virus of Faith'. I hope you'll enjoy the parts of this book where Richard Dawkins hits them for six and then their own so-called "good-book" follows through by kicking them while they're down.

The quotes that I have used may read somewhat differently; (if you check). Please consider the overall meaning, rather than concentrating on individual words. Again I stress, rate your ideas from your own opinion of the text, not from my thoughts or feelings. If your view differs from mine, then there's a chance that you are correct. I cannot possibly be one hundred percent accurate in all things.

CHAPTER ONE

THE WORK BEGINS

At the beginning of his work, Richard Dawkins introduces the theme of—**the elephant in the room**—yet this is easily missed. It's easy to pass-by those words without even noticing; perhaps in part due to the way we process information. For me, *the elephant in the room* becomes apparent after a second look. He then continues on with; "in Britain our government wants to restrict our freedom to criticize religion. Science, we are told should not tread on the toes of theology, but why should scientists tip toe respectfully away? The time has come for people of reason to say; enough is enough, religious faith discourages independent thought, it's divisive and dangerous."

After the brief introduction—the title page, then the main story begins, with the mysterious world of religion. "If you want to experience the **medieval rituals** of faith, the candlelight, incense, music, important-sounding dead languages—nobody does it better than the Catholics. And if we can retain our faith against the evidence, in the teeth of reality, the more virtuous we are."

Now is the opportunity for someone to say, "We should all take a look at (*their*) Book, to see where it actually stands on these strange practices and **medieval rituals**." I commence with our world's well-known and much loved, annual tree-ceremony.

Starting with Jeremiah: Chapter 10 and verse 2. "**Learn not the way of the heathen**, and be not dismayed at the (signs of heaven); for the heathen are dismayed at them, for the customs of the people are vain: for one cutteth a tree out of the forest, the work of the hands of the workman, (with the axe). They deck it with silver and with gold; they fasten it; that it move not." (KJV)

This is describing one of the **ancient** pagan customs, a ritual that was around *before* the modern churches and was later adopted by them. Disguised as their "most popular holiday" the tree-ceremony is still practiced each year by the faithful followers (and non-followers alike). The "signs in the heavens"—that were frightening the superstitious— were in fact **the onset of the winter solstice**, (the shortening of the hours; of sunlight). In *their mind* something needed to be done, to bring back the sun, and make their world safe from harm. The Instruction Book **clearly** informs the church *and the devoted followers* that **such practices are worthless**. This "Tree Ceremony" also includes a "shady-character" who hangs around shopping malls each year. (He appears again in one of the concluding chapters).

CHAPTER TWO

MARY SLEEPS

The following section of the video takes a look at the wonder associated with Mary. The video goes on to show the faithful followers gathering (at a "religious" site). (4:42) "Daylight reveals more of this shrine, where a myth is perpetuated that a virgin who gave birth, (Christ's mother Mary), appeared here, once, to an impressionable young girl. The faithful make the pilgrimage here because they believe that terrible afflictions can be cured, by dragging their poor bodies up to a pool of water, (where Mary made her miraculous appearance)." Then follows; "Religion is about turning untested belief into unshakeable truth through the power of institutions and the passage of time. Catholics believe that Mary was so important that she didn't physically die, instead, her body shot-off into heaven when her life came to a natural end, of course there's no evidence for this."

Here we arrive at his most important point—and the very thing that most people would overlook. These words should grab your attention. This is the part where he points out to viewers that there is a huge discrepancy between The Book, (which the religions claim to follow), and their actual religious beliefs! "**Even the bible says nothing about the way Mary died.** The belief that her body was lifted into heaven

3

emerged after the bible was completed. Made up, like any tale and spread by word of mouth, but it became established tradition."

He then goes on to explain how that tradition became a decree that Roman Catholics must now believe. To consider this, the following questions might be asked.

Was Mary a vital part of their religion?

Did Mary float on up to heaven when she died?

Is it appropriate to pray to Mary as a mediator?

The Book provides some simple answers (as follows). Bear in mind, what is recorded in the Book cannot be later removed or disregarded, cannot be 'back dated' to suit some man-made ideal.

Acts: 1:14. These all continued with one accord in prayer and supplication, with the women, and Mary the mother of Jesus, and with his brothers (KJV)—In this verse we see that they weren't praying to Mary, she was their equal. Also—Matthew: 2:11. And when they were come into the house, they saw the young child with Mary his mother, and fell down, and worshipped—Him. (KJV).

Job 14:10. But man dies, and wastes away: man gives up the ghost, and where *is* he? *As* the waters fail from the sea, and the flood decays and dries up: So man lies down, and rises not: they shall not awake, (not raised out of their sleep). This statement from the past leaves no doubt. Also; (future)—I Thessalonians 4:16. For the Lord himself shall descend from heaven with a shout, with the voice of the archangel, and with the trump of God: and the dead in Christ shall rise first: (KJV).

Also; 1 Corinthians 15:22; For as in Adam all die, even so in Christ shall all be made alive. But every man in his own order: Christ the firstfruits; afterward they that are Christ's *at his coming*.

From these quotes it's easy to see that Mary would be better-off sleeping, like all the others, rather than ending up in the wrong-place, at the wrong time!

I Timothy 2:5; for; there is one God and one mediator between God and mankind (KJV). If the Book claims a total of one mediator then what can they do. How can they make that sentence go away?

Almost everyone has heard of "The Lord's Prayer." In the first few verses it plainly states the words "on earth" (as it is in heaven). The original Mary loved to read the book; she would have known that verse. Hoping to see God's kingdom, "on earth;" (at some future time). Mary would not have wanted to be floating around in space, or in some other dimension if she was praying; "Thy Kingdom come, thy will be done, 'on earth.'"

Another serious problem (for the truth) is the timing of the events. We can see that there was a time when there was nothing special about Mary; then a time when she was believed to have floated on up to heaven; then a time when that news became "official." So, would you please consider the (imaginary) conversation between (the then) self-made pope and the author of the original Book—Something like this perhaps; "Well we sort of told everyone; that you will forget all that 'sleep in the dust' stuff, and go dig up Mary's bones and have her floated-on-up-to-heaven. Oh, by the way, could you back-date the float-on-up-to-heaven part of the process (to suit the fairy story we made up)."

If we can get past such obvious nonsense (from the church) and accept the idea that Mary sleeps to this day; then the question arises, "Who is this Catholic—Mary?" or "Where did she come from?"

CHAPTER THREE

THE EARTH BECAME VOID

The next section (of Dawkins' work) focuses on the age of the earth itself. Comparing religious beliefs with accepted science; such as the fossil record. You might be inclined to see this as simply another 'advertisement for evolution'. Please think back to the introduction, this work is not about promoting 'the theory of evolution'; it's about 'the elephant in the room' (false religion). Another point I would like to make, is that evolution is not under any threat from religion, none whatsoever, as you can see from the following quote. "Even if all the data point to a designer, such an hypothesis is excluded from science because it is not naturalistic." [Todd S.C. (6752): 423, 30th Sept. 1999].

The point being made (by Dawkins) is that, the religions of this world promote the idea of a 'young earth'—an impossible place created instantly—and only a few thousand years ago, which of course flies in the face of scientific evidence. Dawkins points out that the fundamental religionists believe that the book of Genesis describes a designer—who fashioned the earth in just six days. I hope the reader will easily see that this religious theme is just another counterfeit.

To complete this chapter properly, I would need to commence with a great deal of material taken from different sections of the Book, (here a little—there a little, as instructed). I would also need to delve into various definitions of words and phrases, however with that approach you may lose interest. For the benefit of the reader I have taken a step back from the detail, to give a plain statement of what the first verses actually portray. I have then taken another step back to give the all–important *timing* of these events.

You should consider that the earlier versions of the Book (before the translation to English), would have been much more obvious in their meaning, certain words and meanings were 'lost in translation.' The workers already had a world-view; (they 'knew' what it was all about), and they would have been under pressure to 'get-it-right,' (whatever 'right' may happen to be). In some cases getting it right was the very thing that led to errors. An example of unwanted changes comes from the NIV® text, where in two places the word "eagles" was changed to "vultures," perhaps because it was near the word carcass (actually body), however the original word is eagles, the translation was reviewed by fifty scholars and still these errors remain. When you read on from the word eagles, a few verses later in Matthew, the reason for the use of *eagles* becomes obvious. In Luke, if you read the earlier section you can see that, He was talking about 'one being taken and the other left.' Then He was asked; "where." In the answer given, the Hebrew word (soma) was used; def: *"the body" (as a sound whole), used in a very wide application.* Properly translated the verse in Luke 17; is plain "where the whole body of living humans are gathered the eagles will also gather." see: Matthew; 24:28, Luke: 17:37. Note; the word "eagles" appears more than thirty times throughout the Book.

The first verse (page one of Genesis) and the second verse are about two different eras. **Neither** of them refers to the beginning of time (the big bang). If you would like to understand more about the beginning of time; I can recommend "A brief history of time" by Stephen Hawking, it's a terrific story (and easy to follow).

The first verse refers to an era that began approx. three billion years ago! **The second verse** refers to another era that began only six thousand years ago. You may be thinking—"how on earth could he work that out?" Well I say to you, read the information I have provided and then think—"how on earth could it mean anything else?"

Verse one: The beginning of generating an atmosphere **for the first time**. (See note about "shawmayim" *the arch in which the clouds move*).

Verse two: A time much later when the earth had decayed into a state of chaos, covered by an atmosphere (that was fouled); and spinning out of control, a time when the whole earth was covered in darkness.

The book doesn't say, verse one—then a gap of three billion years—then verse two, however, getting these two verses right is the only way the rest of the book can possibly make sense. There is much more to be explained. Try to keep in mind at this point that I am **not** trying to prove that the Book is some kind of science text-book, it is not. I am simply trying to show that anyone who takes a simple approach—without any preconceived ideas—can find a different meaning to what they have been told; (for all those years).

The key to verse one is found by looking up the original Hebrew word for '**the heavens**' shamayim {shaw-mah'-yim}'dual' of an unused 'singular' shameh {shaw-meh'}; *from an unused root meaning to be lofty; the sky (as aloft; the dual perhaps* alluding to **the visible arch in which the clouds move**, as well as to the higher ether where the celestial bodies **revolve**). Popular opinion has all people from BC labeled as "flat-earth" people; yet the Hebrew language envisaged 'a higher ether' as well as revolving celestial bodies.

For verse two, the key to understanding rests with simple words like "was" and "void" and so on. I'll start with void (because that's the easy one). Many readers over the years have assumed that 'void' meant non-existent, but you can see that it really means 'used-up.' One example of this would be the old thick-cardboard railway tickets that were used for country rail travel. When you reached the final destination an attendant would punch-out the word 'VOID,' (so that it couldn't be used again). As you can see, that did not mean that the ticket had not existed in the past, merely that it was now unusable! Now for the slightly more difficult "was", The Hebrew word used is "hayah {haw-yaw}" *to exist, i.e. to become or come to pass*. So if we show the verse differently (using

8

original meaning) we see, 'and the earth **became**'—in place of—'and the earth was,' very little effort is required to make sense of the original text as it was intended.

Now for slightly more difficult—And 'the spirit of God' moved upon the face of the waters. Well what exactly is "the spirit of God." In this case it is simply referring to a being or 'something' that moves along—like a 'probe' or a 'messenger craft,' Humans will send probes to the moon to check the shape of the moon surface, in particular to find out if one 'side' of the moon is flattened. Humans have sent a 'messenger craft' to the planet Mercury, (to take photos); posted on NASA's 'Messenger' web site. So as you can see, something "moving upon the face of the waters" may have seemed a bit far-fetched (to previous generations), by today's standards however; with a steady diet of science-fiction movies combined with real images from space; a being or craft from another dimension is not that hard to contemplate.

There must be more evidence to show that the first verse describes a different 'era' from the rest of page one. Understanding the remainder of page one hinges on a simple use of the word 'after' so it should be easy, but first some obvious differences between the two eras. The start of the book of John goes back even further (before verse one) so that's a good place to begin the explanation. John also explains that there were two beings, which should be a wake-up for those learning about a triune; or a 'trinity' of beings. John 1:1. In the beginning was the Word and the Word was with God, and the Word was God. The same was in the beginning with God; all-things were made by Him; and without him was not anything made *(KJV)* Now those few sentences can cause some confusion, but it doesn't have to. The name used is a **family** name. For example suppose a man's surname was Smith and his son's name was Bill Smith and the two of them designed and built a device; (while working together as a family). You would be able to say—"At the beginning (of the project); was Bill; and Bill was with Smith; and Bill was Smith; and he was with Smith from the beginning.

For two beings to design and release a three dimensional universe they would need to already have existed in another dimension, (perhaps the fifth or sixth dimension). We may contemplate (but never actually see); the second dimension, and since we have dibs on the third and fourth; (i.e. we do 'time' calculations on the movement of stars). In other words; they would have already existed in a place where time (as we know it) was not that critical, (mainly because it hadn't started yet).

Now; for some detail about verse one. The following is a question asked of a man named Job, it seems at first to be confusing; and even a tad far-fetched, until someone is willing to apply some critical thinking. Job 38:5. Question; "Who hath laid the measures thereof, if you know? Or who hath stretched the line upon it? Whereupon are the foundations thereof fastened? Or who laid the corner stone. When the morning stars sang together, and all the sons of God shouted for joy?" (KJV). This is a question put to Job in a language that he could understand, but Job knew it was not about a building site. It was in fact a question about the earth itself. Now the last part of the question doesn't seem to mean much, and yet there are three critical points to consider.

These 'stars' or 'angels' (millions of them) were individually created beings, (not male or female like Man-kind), and they were already in existence **before** the earth was made.

They were **all** happy and in a state of **co-operation**, so there was no deception at that time, no liars.

The earth must have been a **finished product** and must have been pleasing to look at, so they would all shout for joy. In short, the earth must have been made beautiful and ready to occupy, (not at all like the one described in verse two-which was billions of years later.)

Of course this cannot be referring to the time of the 'big bang', as the earth would have been a tiny spec, inside a condensed ball of Energy.

What about the possibility of someone 'making' angel-beings? We need to stop and consider just how far-fetched that claim might be. I can imagine in early times, superstitious people simply wouldn't have had a chance of making sense of all that. But what about us, what about today's generation? Mankind can produce computer hardware each with its own serial number. We can produce various kinds of computer languages and software, each with its own serial number. We can 'pump out' these products by the millions, and think nothing of it. If you can do it, try to imagine a book of the future, say three billion years from now, the book is making a claim about the millions of computers that mankind designed and produced. Would the reader of the future find such a task hard to believe? Yet we **have** achieved such a thing, (and we are mere humans living in the third dimension). Of course there is much more to discover, but I believe I have shown enough to distinguish between the two ages. Before verse two; all the angels were in a state of co-operation. Between the verses—one third of them (the ones living on earth)—changed into a permanent state of competition; (opposition to their manufacturer). Before verse two the planet was in a beautiful state—ready to occupy. By the time of verse two, the planet was in a state of chaos and darkness.

Well, you can see by now, that it would take 'many books' to cover the events that might have occurred in that tiny space (just a small gap between two verses). You can see that the real 'Instruction Book' could not possibly be twisted and distorted into supporting a young earth. If your religion is teaching a creation in six days, you know they haven't been reading the same book that I have.

The rest of the early pages describe the environmental recovery of the planet, (bringing it back from a state of chaos). When anyone sees the word 'after' in general use it nearly always has a 'before' to go with it. For example: If I say "I will see you after lunch," you can assume the time of speaking is 'before' lunch, and so on. The 'making' of plants and animals described in the early parts of Genesis, is actually a 're-making' of 'kinds' that had already existed on the earth (from a previous era). This is a claim of DNA technology. This is a claim of already understanding that a plant's own DNA is a blueprint for life. This re-making would seem far-fetched to anyone reading the book—say two hundred years ago—but what about today?

To give a comparison to human technology, it would be similar to storing samples of male and female DNA (usually in the form of blood) in a cryogenics lab, in the hope that some future scientist might be able to 're-make' the original 'kind' of creature. This form of technology is already available, and is being put to use by humans at present, **even though we haven't yet perfected the recovery stage**. The rest of it is fairly basic stuff. Once you discover that the word 'heavens' means either the atmosphere or; (in some cases); the universe; Then the rest of Genesis begins to make a whole lot of sense. The light became visible when the thick dark atmosphere was cleared; (obviously an environmental improvement). The dividing of night from day would have required the return of the earth's rotation to a twenty-four hour period; and so on. Such an effort is really not that hard to believe. You can be sure that if future scientists, find a planet similar to earth, they will try to figure out how to change certain aspects, to create a more earth-like environment.

CHAPTER FOUR

GO ONE FURTHER

The remainder of the first video covers some interesting stuff, and makes some good points about religious practices. However it mostly falls outside the area of the work. For the most part I will be using the fast-forward button. I really should ignore pastor Ted, but I must show one thing. He was teaching the faithful followers that they were 'set free' from sin. I would simply say to him, "please take a look at the 'book of Romans' before you start waffling on." But what exactly is sin? Well it simply means—breaking one or more of the ten basic rules, (or laws) which are set in place. 1 John 3:4; ". . . for sin is the transgression of the law." There are rules about stealing or killing, rules about how to treat other people, even a rule about taking a day off. Some rules are about 'what to do,' some are things 'not to do.' There are only ten, so it should be fairly easy, right? Well apparently not! Romans: 3:23. For all have sinned, and come short of the glory of God; (KJV). Romans 8:7; the carnal mind is enmity against God; for it is *not* subject to the laws of God, neither indeed can be (KJV). Romans 6:23; the wages of sin *is* death (KJV). I really don't think pastor Ted would be saying that his followers are set free, if he had actually read the Book as he claims.

The following section of video is about the continuing struggle between one form of religion or another. This is where I "switch off," so please zoom forward to the point (44 minutes) where Dawkins warns about; "believing because you've been told to believe, rather than believing because you've looked at the evidence." Surely that deserves a supporting comment from the book?

1 Thes: 5:21. Prove all things; hold fast that which is good (KJV).

Nearing the end of the first video we have the following statement. "We are all atheists about *most* of the gods that societies have ever believed in, some of us just go one god further."

That is more relevant than you could imagine. When I cover some of the present-day customs; in another section; you might recall the statement above. (You might even want to remove a goddess or two from your own life).

CHAPTER FIVE

SEGREGATION

At the start of the second video ('The virus of faith') we are taken through a dialogue about the separation of young children into groups—which are then taught a particular 'brand' of religion. We are then taken through a question and answer scenario with a Jewish leader.

Richard Dawkins: "In this program I want to examine two further problems with religion. I believe it can lead to a warped and inflexible morality, and I'm very concerned about the religious indoctrination of children. I want to show how 'faith' acts like a virus that attacks the young and infects generation after generation."

"I want to ask whether **ancient mythology** should be taught in schools. It's time to question the abuse of childhood innocence with superstitious ideas of hell-fire and damnation. There is something exceedingly odd about the idea of sectarian religious schools. If we hadn't got used to it, over the centuries, we'd find it downright bizarre."

"When you think about it, isn't it weird the way we automatically 'label' a tiny child with its parents religion. Nobody would categorize children by the political party; which the parents support. We agree they're too young to know where they stand on questions of politics, so why is it not the same for where they stand on the cosmos, and humanity's place in it?"

"Children are initially separated from each other because of their parents faith, then their differences are constantly drilled into them, and they embark on opposing life trajectories. Such divisions are encouraged—not only in Israel, but right on [our] doorstep, in Northern Ireland for instance, or in London."

"In North London the Hasidic Community is the largest after Israel and New York. Here, religious division is taken to its extreme. Television is frowned upon, and of course children attend exclusive religious schools, clustered away from external influences—which just might persuade them to look outside their community. I want to find out why these children are being segregated, and whether their culture allows them to open their minds to reality."

A message to the teachers of 'Judah' can be found in Matthew 23; "The Scribes and the Pharisees have the authority of Moses. All things then which they teach [from the book], do these **and keep**, but do not take their 'works' as your example, for they say—and do not. They make **hard laws** and put great weights on men's backs; **but they themselves will not put a finger to the task**. (KJV).

If you can see the 'context,' (their works) represents the burdensome traditions; or; the 'man-made' parts of their religion. Also in; Luke 13:34; O Jerusalem, Jerusalem, which killest the prophets, and stonest them that are sent unto thee; how often would I have gathered thy children together, as a hen doth gather her brood under her wings, and ye would not. In short, He wanted Judah to change, to *become* obedient.

Dawkins: "We live in the shadow of a religiously inspired terror, in an era when science has plainly shown religious superstitions to be **false**. And yet it's a strange anomaly that 'faith-schools' are increasing in number and influence **in our education system**, with active encouragement from [Tony Blair's] government. There are already seven thousand faith schools in Britain and the government's trust reforms are encouraging many more. Over half the new city academies; are expected to be sponsored by religious organizations. The most worrying development is a new wave of private evangelical schools that have adopted the American Baptist 'A.C.E.' curriculum (Accelerated *christian* Education)." In the interview, this guy (Adrian) stumbled through the idea of the earth being 'made' in six days. At least he was honest and said "I don't know," and then he went on to say, "it's a sort of an academic question really, and I don't care about the answer very much really. Does that make sense?" Do I even need to waste a comment on such a person?

Dawkins: "But why should he impose his personal version of reality on children? Not only are they encouraged to consider the weird claims of the Bible; alongside scientific fact; they are also being indoctrinated into what an objective observer might see as a warped morality." Here we see Dawkins associate 'the weird claims' with the Bible, however you should realise that he has entered into this with his own view of how religion works. Richard Dawkins is kept busy with his own work as an evolutionary scientist, he simply doesn't have the time to consider the consequences; of the vast disconnect that has occurred over the centuries (even though he has hinted at it many times throughout his work). The **correct** wording for the sentence above is as follows. [Not only are they encouraged to consider the weird claims of **false religion**—alongside scientific fact, they are also being indoctrinated into what an objective observer might see as a warped morality.]

Dawkins: "Let me explain why, when it comes to children, I think of religion as a dangerous virus. It's a virus, which is transmitted partly through 'teachers' and clergy, but also down the generations from parent to child to grandchild. Children are especially vulnerable to infection by the virus of religion. A child is genetically pre-programmed to accumulate knowledge from figures of authority. The child brain, for very good reasons, has to be 'set up' in such a way that it believes what it's told by its elders, because there just isn't time to experiment with warnings, like "don't go too near the cliff edge," or "don't swim in the river, there are crocodiles!" Any child who applied a scientific (questioning) attitude would be dead. No wonder the Jesuit said, 'give me the child for his first seven years, and I'll give you the man.' The child brain will automatically believe what it's told—even if what it's told is nonsense. And then when the child grows up, it will tend to pass on that same nonsense to its children."

CHAPTER SIX

THE DARK PLACES

Dawkins: "For many people, part of 'growing up' is killing off the virus of faith, with a good strong dose of rational thinking. But if an individual doesn't succeed in shaking it off, his mind is stuck in a permanent state of infancy, and there is a real danger he will infect the next generation. I'm going to meet someone who has experienced 'religion as child abuse,' first hand. Jill Mytton was brought up in a strict 'christian' sect. Today she's a psychologist who rehabilitates young adults, similarly scarred by their narrow religious upbringing."

Jill Mytton: (London Metropolitan University); "They need to be allowed to hear different perspectives on things. They need to be allowed to investigate. They need to be allowed to develop their critical faculties, so that they can take a number of different viewpoints and weigh them up, and decide which one is for them. They need to find their own pathway, not to be forced into a particular mold as a child. If I think back to my childhood, it's one that's kind of dominated by fear, and it was a fear of disapproval, while in the present, but also of eternal damnation. To a child, images of hell-fire and gnashing of teeth are actually very real and not metaphorical at all. If you bring a child up and discourage it from thinking freely, and making choices freely, then that's still (to me) that is a form of mental abuse, or psychological abuse." Then, after a question on [hell], "It's strange isn't it, after all this time, it still has the power to effect me. [Hell] is a fearful place with torment and torture and it goes on forever, where there is no respite from it."

Dawkins: "It's deeply disturbing to think that there are 'believers' out there who actively **use** the idea of [hell] for moral policing. In the United States, 'christian' obsession with sin has spawned a national craze for 'hell-houses' (morality plays—come halloween freak shows) in which the evangelical 'hobby-horses' of abortion and homosexuality are literally demonized. Pastor Roberts is rehearsing a new production for his Colorado based hell-house, which he has written and staged for almost fifteen years. He fervently believes that you have to scare people into being good."

Pastor Keenan Roberts: "To call upon my life as a pastor, as a minister, *is to tell people what the book says*. And what I, and we, and our church, and hundreds of churches across this country, and around the world are doing is, we have found a very creative, effective tool that is getting people's attention, to consider the message. We want to leave an indelible impression upon their life, that sin destroys."

Meanwhile a message flashes on screen, which shows completely the opposite of what he believes. Wouldn't you think that after fifteen years of reading, he would have had time to consider, what that message plainly states? The message onscreen shows (in plain English) "The wages of sin is death." No amount of twisting and distorting, by these wannabe ministers, can turn that plain statement into something supporting their own personal 'invention;' known as 'false religion' or 'churchianity. Death means death—not torment.

At this point we need to clear up a few things about exactly what 'hell' is. What is the true definition, and what does the Instruction Book say, (about where people go when they die). Our English dictionary has all sorts of weird and wonderful ideas about "hell." However, such definitions come from (our) world-view. To see the true picture we need to look at the original word, as it was intended, at the *time* it was used, (in the Instruction Book). That's really the only way to find the true meaning of the word. The word used in the Hebrew language was—showl; {sheh-ole}; which is defined as grave or pit.

Now let's look at a whole sentence that might have confused a few ignorant, superstitious ministers in the past. Psalms: 16:10; thou wilt–not leave my soul in hell; neither wilt thou suffer thine holy one to see corruption. Here we see the words 'hell' and 'suffer' in the same sentence. Now let's take a look at the 'plain-English' translation of the above, starting with the required Hebrew words and definitions. You should be ok with the words . . . (For you will not leave my . . .)

Soul—Hebrew: nephesh {neh'-fesh} properly, a breathing creature, i.e. animal of (abstractly) vitality; used very widely in a literal, accommodated or figurative sense (bodily or mental) any, appetite, beast, **body, breath, creature.**

Hell—Hebrew: showl (sheh-ole} the **grave** or pit.

Suffer—English (8); allow something: to allow something to happen or to be done (archaic or literary); example: Suffer the little children to come unto me. [Encarta ® World English © Microsoft (: o)

Corruption: Hebrew: shachath {shakh'-ath} a pit (especially as a trap); figuratively **destruction.**

(For thou wilt not leave my soul in hell; neither wilt thou suffer thine holy one to see corruption). Now if we strip away the religious nonsense, we can plainly see the written meaning in the text. [For you will not leave my body in the grave; neither will you allow your holy one to see final destruction]. And finally in my own (Australian) language it would simply say . . . "If I cark-it, and my body gets dropped six foot under, then please bring me back to life (at the end of the era); and I really don't want to be killed off a second time, in that place of final destruction!"

The **false** idea of suffering in 'hell,' follows on from the **false** idea of an immortal soul. The Instruction Book plainly states that man became a living soul. It definitely does **not** say, man has an *Immortal* soul'. The idea of an *immortal* soul must have been added by churchianity. Perhaps they thought it would be good for business, to offer the customers a cushy life in heaven, (in return for supporting the church). The idea certainly did not come from the Book. So it is plain to see, that without an *immortal* soul, you're not going to suffer in hell and you're not going to 'float on up to heaven' either, you are simply going to sleep; as clearly stated so many times throughout.

John: 11:11. These–things said he: and after that he saith unto them, our friend Lazarus sleepeth; but I go, that I may awake him out of **sleep**. Then said his disciples, Lord, if he sleep, he shall do well. Howbeit Jesus spake of his death: but they thought that he had spoken of taking of rest in sleep.

Daniel: 12:2; And many of them that **sleep** in the dust of the earth **shall** awake, some to everlasting life, and some to shame . . . Notice it states they shall—(it hasn't happened yet)

Psalms: 13:3; . . . lighten mine eyes, lest I sleep the sleep of death (KJV).

CHAPTER SEVEN

MURDER

At the start of this work, I was determined not to get involved with the attacks, on the contents of the Book. However, when the matter of 'stoning people to death' was raised I realised that it was linked to a well-known saying, something that even non-readers would recall. The story is told in the New Covenant and it should clear up any doubt. It is included for those people who think they can make their own judgement, then carry out the sentence 'by the book.' The saying is . . . "Let he who is without sin—cast the first stone."

John; chapter 8: And early in the morning he came again into the temple, and all the people came unto him; and he sat down, and taught them. And the Scribes and Pharisees brought unto him a woman taken in adultery; and when they had set her in the midst, they say unto him, 'Master, this woman was taken in adultery, in the very act. Now Moses in the law commanded us, that such should be stoned: but what sayest thou?' This they said, tempting him that they might have—to accuse him. But Jesus stooped-down, and with *his* finger wrote on the ground, *as though he heard them not.* So when they continued asking him, he lifted up himself, and said unto them, "**He that is without sin among you—let him first cast a stone at he**r." And again he stooped down, and wrote on the ground. And they which heard *it*, being convicted, *by* *their own conscience*, went out one by one, beginning at the eldest *even* unto the last: and Jesus was left alone, and the woman standing in the midst. When Jesus had lifted up himself, and saw none but the woman, he said unto her, Woman, where are those thine accusers?

Hath no man condemned thee? She said, 'No-man, Lord.' And Jesus said unto her, neither do I condemn thee: go, and sin no more. (KJV)

Dawkins:(31 minute) on the second video; "If you take the good book to its extreme (and some people do) you can justify murder. In 1994, Paul Hill shot and killed Dr. John Britain outside his abortion clinic in Florida. In 2003 Hill was executed for murder, but he went to his death, claiming his actions were backed by holy-scripture. I'm going to meet Paul Hill's friend and defender, Michael Bray. On what moral basis could he, as a 'christian,' defend a self-professed, cold-blooded killer?"

Dawkins: "You're friend Paul Hill, who was convicted of murdering a doctor, he took the law into his own hands, didn't he?"

Michael Bray: "No, Paul Hill by his own testimony acted defensively. Not in retribution, that's the job of the law, the job of the law is to punish. The job of citizens is indeed (out of love) to protect one another."

Well have you ever heard such a terrible load of rubbish, and this man is claiming to be a minister. By the way, a genuine minister would know not to call himself 'reverend.' If this man learned to read instead of soaking up 'waffle'—he might manage to learn right from wrong.

Luke: 6:37: Judge not, and you shall not be judged: condemn not, and you will not be condemned: forgive, and you will be forgiven. (KJV)

This is not about whether the doctor was right or wrong, (in his actions). It is a question of who has the 'right' to judge the doctor for his actions. To put it simply, murdering another person cannot be justified, not by man's laws and certainly not by using a Book to justify that action. The same rule applies to murdering yourself, this is a message that is often used successfully, when talking to someone who is about to do something silly (like jump off a bridge). The person might say

"I have all these terrible problems, I can't seem to sort things out, I'm better off just ending it all now." Then a helper can quietly say to them, "what if someone, one day, figures out how to bring your body back to life. Then you may still have to face all your terrible problems—and a self-murder charge to go with it." For the brief summary that follows, I would like you to slip in the word '**false**' before the word 'religion,' then the message becomes much clearer.

Dawkins: "It was curious, I quite liked him, I thought he was sincere, I thought he wasn't really an evil person. I was reminded of a quotation by the famous American physicist; 'Stephen Weinburg' (Nobel Prize winning theoretical physicist). Weinburg said; "Religion is an insult to human dignity—without it, you'd have good people doing good things, and evil people doing evil things—But for good people to do evil things, it takes (—) religion."

CHAPTER EIGHT

WINDOWS

Dawkins: "I want to examine how science reveals the true roots of human morality. Morality stems, not from some fictional Deity and His text, but from altruistic genes that have been naturally selected in our 'evolutionary' past."

Oliver Curry: (London School of Economics); "Humans have much more sophisticated versions of the kind of social instincts you see in chimps and other creatures—really there's no great leap; its just, if you can think of chimps as 'ms-dos' and humans as 'Windows 2000.'

Well, as I said at the start, I don't want to get into a debate over evolution, but if you're a science teacher, and your using that example as part of a lesson, you should set the students an extra question for homework.

(Since this is comparing forward progress)—What did the 'ms-dos' computer have going for it, that the chimp couldn't possibly have had? I would suggest offering an incentive—say a few days without homework, (to ensure you receive some intelligent replies).

I am finishing off the first section with a message from Ian McEwan, followed by an appropriate closing statement by Richard Dawkins. And, for those readers who think religion will; one-day; be booted out of our education system. After that; if someone wants to investigate "The Book," please let it be their choice.

Ian McEwan: "I guess my starting point would be, the brain is responsible for consciousness, and we could be reasonably sure that when that brain ceases to be, when it falls apart and decomposes, that'll be the end of it. From that point a lot of things follow, I think, especially morally.

We are the very privileged owners of a brief spark of consciousness, and we therefore have to take responsibility for it. You cannot rely on a world elsewhere—a paradise to which one can work towards, and maybe make sacrifices, and crucially make sacrifices of other people. We have a marvelous gift, and you see it develop in children—this ability to become aware that other people have minds, just like your own, and feelings that are just as important as your own. And this **gift** of empathy, seems to me to be the building block of our moral system . . ."

Richard Dawkins: "Look around you, nature demands our attention, begs us to explore, to question. [—] Religion can provide only facile, unsatisfying answers. Science; in constantly seeking real explanations, reveals the true majesty of our world in all its complexity. People sometimes say; there must be more than just this world, than just this life—but how much more do you want? We are going to die, and that makes us the lucky ones. Most people are never going to die, because they're never going to be born. The number of people who could be here in my place, out-number the sand grains of Sahara. If you think about all the different ways in which our genes could be permuted, you and I are quite grotesquely lucky to be here. The number of events that had-to-happen, in order for you to exist, in order for me to exist. We are privileged to be alive, and we should make the most of our time on this world."

Presented by: Professor Richard Dawkins
          University of Oxford

CHAPTER NINE

THE REST

"The Rest" has a double meaning. On the one hand, this chapter is the beginning of the rest of the book. It is also a story about entering "The Rest," (i.e. obeying the law).

What on earth could possibly drive someone, anyone, to take on a desire to remove the first day of the week, and shove it all the way down into eighth place, and then push all seven days back one place; deceptively re-engineering the working week? Well that is exactly what happened when Pope Gregory decided to invent a new calendar (the Gregorian calendar), in order to force people to "remember" the new seventh day of the week. When you think about it, making a new week would have been a lot simpler than recalling the books and destroying them, in order to maintain the new 'traditional' beliefs. He would have been bitterly disappointed, when he realised that his efforts were in vain. The Book still tells us which day is which, and nowadays we are privileged to be able to read for ourselves (rather than being told mysterious things in foreign languages). There can be no doubt whatsoever, that Sunday is the first day of the week. This chapter explains a little about whether Sunday is a good day, or should the Sabbath day be kept (technically sundown Friday to sundown Saturday). The Sunday problem also gets a mention in the

'Ishtar' chapter. Also the calendars are briefly mentioned in relation to 'Ephraim and Manasseh' (in a later chapter).

Firstly, I need to explain how the Roman leaders gave themselves the **"authority"** to change the day (i.e. to suit their own traditions). Their **supposed** authority comes from Matthew: 16:18; And I say also unto you, that you are Peter. And upon this rock; I will build my church; and the gates of hell shall not prevail against it. And I will give unto you the keys of the kingdom of heaven: and whatsoever you shall bind on earth shall be bound in heaven: and whatsoever you shall loose on earth shall be loosed in heaven (KJV).

Please note . . . 1 Corinthians 3:11. For **other** foundation can no man lay than that is laid, which is Jesus Christ (KJV).

It's easy to see how someone could become confused by some verses, however the leaders at the time, would have been reading this in the Greek language, the true meaning (then) would have been much clearer. The deception applied to these verses was no accident! By referring to the Greek definition the difference between the two 'rocks' is obvious. The word translated 'Peter', is from the Greek word 'Petros' apparently a primary word; a piece of rock. So you can see 'Peter' was a 'piece of rock' (a little bigger than a small stone). That's exactly how it would have read to them—all those years ago. The second 'Rock' is translated from the word 'Petra' meaning a mass of rock (literally or figuratively): rock. So we end up with—"You are Peter," (followed by referring to himself). "And upon **this rock** I will build my church. **I** might add; (And even if someone tries to kill-off all the believers, the church will still prevail); God's church could still continue as a 'little flock'; even to this day. Now the part following that, is supposedly the 'authority' to change whatever you like. You can see that; even if this 'invented' church were allowed to bind anything, it would have to be within the boundaries of the rules and laws clearly given. Please notice in the following, the text says **will** do; (at some **future** time). Let's continue anyway. (And I will give unto you the keys of the kingdom of heaven: and whatsoever you shall bind on earth shall be bound in heaven). Notice that it is written clearly "I will give,"(future). Notice also that the words 'of heaven' are clear. This is definitely not referring to Peter being in heaven (at that future time). Peter will be given the keys to the kingdom [from] heaven and will then have authority (and hopefully the wisdom) to bind and loose, (according to the laws of God). He most certainly was not given authority at the time referred to in the book of Matthew. If you doubt it, simply skip forward to verse 23: (. . . He turned, and said unto Peter, Get behind me, Satan: **you are an offence to me**: for you savour not the things that be of God, but those that be of men) (KJV).

So the more you look into it, the more you realise that the supposed authority, of this "man-made church" was never more than an **elaborate hoax**. The man (Peter) was only human after all.

Some will say that the present seventh day cannot be the same because some days were taken out of the month in the year 1582 (to make an adjustment). No, actually only the 'dates' were changed; Thursday, October 4th, was followed by Friday, October 15th. Now let's see what the 'church' has to say about Sunday, by reading through some of the following quotes . . .

"I have repeatedly offered one thousand dollars to anyone who can prove to me, (by the Bible alone), that I am bound to keep Sunday. There is no such law in the Bible. It is a law of the Holy Catholic Church alone. The bible says 'Remember that you keep holy the Sabbath day. The Catholic Church says, No! By **my** (divine) power I abolish the Sabbath day, and command you to keep holy **the first day of the week**. And lo! The entire civilized world bows down in reverent obedience to the command of the holy Catholic Church." [Bishop Thomas Enright in a personal letter, printed in 'Experiences of a Pioneer Minister of Minnesota' by W.B. Hill]

From Peter Geiermann; "Question: which day is the Sabbath day? Answer: Saturday is the Sabbath day. Question: Why do we observe Sunday instead of Saturday? Answer: We observe Sunday instead of Saturday because the Catholic Church in the council of Laodicea (AD 363), transferred the solemnity to Sunday."

[Peter Geiermann, The Convert's Catechism of Catholic Doctrine].

From a magazine article, printed in 1975:

"1) Protestants, who accept the Bible as the only rule of faith and religion, should by all means go back to the observance of the Sabbath. The fact that they do not, but on the contrary observe Sunday, Stultifies them in the eyes of every thinking man."

"2) We Catholics **do not accept the Bible** as the only rule of faith. Besides the Bible we have the living church, the authority of the Church to guide us. We say, This Church—(instituted by Christ to teach and guide man through life), has the right to change the ceremonial laws of the Old Testament and thence, we accept **her change** of the Sabbath to Sunday. We frankly say, yes, the Church made this change, made this law, as she made many other laws, for instance the Friday abstinence, the unmarried priesthood, the laws concerning mixed marriages, the regulation of Catholic marriages, **and a thousand other laws** . . . It is always somewhat laughable to see the Protestant churches, in pulpit and legislation, demand the observance of Sunday, **of which there's nothing in their Bible**. [Peter R. Kraemer, Catholic Church Extension Magazine, 1975].

Anyone can see from these quotes (and hundreds of others) that The Catholic Church wants the exclusive right to Sunday. They want the daughter churches to **give up on Sunday** and return to God and keep the Sabbath Day as commanded.

Have you ever heard the old saying, 'Manna from Heaven?' The casual reader would not expect to find 'manna' in a story about 'The Rest,' however such a story shows just how important the seventh-day really was. Hebrew: **Manna**; literally, a what-ness (so to speak), i.e. manna (so called from the question about it) = "**What is it?**"

Exodus: 16:13; . . . in the morning the dew lay round about the host. And when the dew that lay was gone up, behold, upon the face of the wilderness there lay a small round thing, as small as the white frost on the ground. And when the children of Israel saw it, they said one to another, It is manna: **for they wist not what it was**. And Moses said unto them. "This is the bread which the Lord hath given you to eat. This is the thing which the Lord hath commanded, Gather of it every man according to his eating, an **omer;"**

[An ancient Hebrew unit of dry measure, roughly 3.5 liters; Encarta® World English © Microsoft].

". . . For every man, according to the number of your persons; take ye every man, for them which are in his tents." And the children of Israel did so, and gathered; some more; some less. And when they did mete it with an omer, he that gathered much had nothing over, and he that gathered little had no lack; they gathered every man according to his eating. And Moses said, Let no man leave of it till the morning. Notwithstanding they hearkened not unto Moses; but some of them left of it until the morning, and it bred worms, and stank: and Moses was wroth with them. And they gathered it every morning, every man according to his eating: and when the sun waxed hot, it melted. And it came to pass, that on the sixth day they gathered twice as much bread, two omers for one man: and all the rulers of the congregation came and told Moses. And he said unto them; Tomorrow is the rest of the holy Sabbath unto the Lord: bake that which ye will bake today, and seethe that ye will seethe; and that which remaineth over, lay up for you to be kept until the morning. So that on the sixth day the 'Tribes of Israel' would bake enough for the seventh also, (they were prepared for the Sabbath). On any other day of the week the bread would not keep until the following day!

Any 'talk' about floating to heaven usually leads to jokes about dead-people seeing St. Peter at "The Pearly Gates." The actual text is quite different from this popular idea. Do I need to include a definition of the word down? As far as I can tell, that's a word that hasn't changed much and should be easy enough to understand. The following is the vision (or daydream); which John published (90 AD).

Revelation 21:2; And I John saw the holy city, new Jerusalem, coming **down** from God out of heaven . . . (KJV)

31

Revelation 21:10; and he carried me away in the spirit to a great and high mountain, and showed me that great city, the holy Jerusalem, **descending** out of heaven . . . (KJV)

Revelation 21:21; and the twelve gates were twelve pearls; every several gate was of one pearl: and the street of the city was pure gold, as if it were transparent glass (KJV).

John's message to the churches is plain. If Peter, in the future, is likely to see the 'pearl gates' it would be on earth; (after the city has descended.)

CHAPTER TEN

EPHRAIM AND MANASSEH

Throughout The Book there is a great deal of information about two brothers; Jacob and Esau. Stepping back from there (a little); we can find the story of Abraham's son; Isaac; and Isaac's wife Rebekah; they were trying to have a child; (a son to carry on the inheritance). This was a period of time when having a son was important; (and a first son would receive a greater portion).

From this point, it's a good idea to skip forward a page or two; and then come back to these definitions or quotes if needed! The following; partly describes where the twelve tribes originated.—

Please consider the definition of the words Jacob and Esau; from the Hebrew language; also the definition of . . .

Bowels: [me'ah may-aw'] from an unused root, meaning to be soft; used only in plural: the intestines, or (collectively) the abdomen, figuratively, sympathy; by implication, a vest; by extension, the stomach, the uterus, the heart (figuratively): belly; **or womb**.

Jacob: [Ya'aqob] {yah-ak-obe'} (i.e. supplanter); Jaakob, patriarch; from 'aqab aw-kab' a primitive root; properly, to swell out or up; to seize by the heel; figuratively, to circumvent (as if tripping up the heels); also to restrain (as if holding by the heel): take by the heel, stay, supplant-utterly. Note; there was no J back then, only the 'sound.'

Esau: 'Esav ay-sawv' apparently a form of the passive participle of 'asah aw-saw' in the original sense of handling-rough (i.e. sensibly felt); the Israelitish Esav, a son of Isaac, including his posterity.

Genesis 25:21; And Isaac intreated the Lord for his wife, because she was barren: and the Lord was intreated of him, and Rebekah his wife conceived. And the children struggled together within her; and she said; "if it be so, why am I thus?" And she went to enquire of the Lord. And the Lord said unto her, "two nations are in thy womb, and two manner of people shall be separated from thy bowels; and the one people shall be stronger than the other; and the elder shall serve the younger" (KJV).

Genesis 25:24; And when her days to be delivered were fulfilled, behold, there were twins in her womb. And the first came out red all over, (like an hairy garment); and they called his name Esau. And after that came his brother came out, and his hand took hold on Esau's heel; and his name was called Jacob: and Isaac was threescore years old (when she bare them). And the boys grew: and Esau was a cunning hunter, a man of the field; and Jacob was a plain man, dwelling in tents.

Genesis 27:36; and he said, "Is not he rightly named Jacob? For he hath supplanted me these two times: he took away my birthright; and, behold, now he hath taken away my blessing." And he said, "hast thou not reserved a blessing for me?"

Genesis 32:28; And He said, "Thy name shall be called no more Jacob, but Israel: for as a prince hast thou power with God and with men, and hast prevailed."

Israel had twelve sons—Reuben, Simeon, Levi, Judah, Zebulon, Issachar, Dan, Gad, Asher, Naphtali, Joseph and Benjamin. These families grew into a nation. Israel lived in Egypt, in the country of Goshen; and they had possessions, and multiplied.

Numbers 2:1; And the LORD spake unto Moses and unto Aaron, saying, every man of the children of Israel shall pitch by his own standard, with the ensign of their father's house: around the 'Tent of Meeting' at a distance. (KJV)

Numbers 13:11; And the Lord spake unto Moses, saying, Send thou men, that they may search the land of Canaan, which I give unto the children of Israel: of every tribe of their fathers shall ye send a man, every one a ruler among them. And Moses by the commandment of the Lord sent them from the wilderness of Paran: all those men were heads of the children of Israel. And these were their names: of the tribe of Reuben, Shammua the son of Zaccur. Of the tribe of Simeon, Shaphat; the son of Hori. Of the tribe of Judah, Caleb the son of Jephunneh. Of the tribe of Issachar, Igal; the son of Joseph. Of the tribe of Ephraim, Oshea; the son of Nun. Of the tribe of Benjamin, Palti; the son of Raphu. Of the tribe of Zebulun, Gaddiel; the son of Sodi. Of the tribe of Joseph, (namely) of the tribe of Manasseh, Gaddi; the son of Susi. Of the tribe of Dan, Ammiel; the son of Gemalli. Of the tribe of Asher, Sethur; the son of Michael. Of the tribe of Naphtali, Nahbi; the son of Vophsi. Of the tribe of Gad, Geuel; the son of Machi. These are the names of the men; which Moses sent to spy out the land. And Moses called Oshea the son of Nun; Jehoshua.

—Their camp eventually became well-organized; each tribe set-up (around the tent of meeting) according to their family name.

To the north: Nephtali, Asher, Dan.
To the south: Rueben, Simeon, Gad.
To the east: Zebulan Issachar, Judah.
To the west: Ephraim, Manasseh, Benjamin.

Now' if you've been paying attention; you would have noticed that Levi has disappeared, and two fresh names are now included. Joseph had two sons; Manasseh and Ephraim; (a nation and a company of nations). Levi was a separate tribe, set-aside for special duties.

The tribe of Dan would move away from Israel (and perhaps give up their right to be included in the promise). Dan traveled in ships and ended up settling in Ireland; (after leaving a trail of places with their father's name). Even though the tribe of Dan took-off at the wrong time, they may have 'paved the way' for later migrations to other countries; (like Scotland and England).

In Australia; I often hear comments like "I don't like Israel," or "I don't like the way Israel behaves." If you are living in Britain, Australia or the United States, you could be naming yourself! Yes, the person saying 'Israel' could in fact be a member of one of the twelve tribes of Israel.

Are you in one of these? Perhaps if you prefer a calendar with Saturday at the end of the week; you could be in Ephraim, on the other hand if you like the Sunday at the end, you could be in Manasseh (or you could have been 'converted' by your place of employment, one example of this is when government departments buy computer software 'offshore'. Notice; firms in Australia will sometimes issue a promotion-calendar thinking they are getting free advertising, year-round. Some people (in Australia) will toss a calendar in the bin, unless the first day is Sunday, (i.e. the calendar week ends on Saturday); so the advertising is wasted.

Do you normally drive on the left-hand side of the road? You are probably 'Ephraim'. If your bathroom has a 'tap' and your car has a 'boot' then you are probably 'Ephraim'. Another giveaway for Ephraim/Manasseh is your spelling. Words used by people, companies (or political parties), who are Ephraim—will have or use the odd extra letter here and there, in their words (compared to Manasseh); including words like colour or, labour; and so on. There is often a certain hidden style in a company; (or party); disclosed by their name or title. For example (Australian Pensioners Insurance Agency) is really a 'Manasseh' company in Australia, note there is nothing wrong with that, they are a suitable insurance company, however the full name would be (The American Insurance Company for Australian Pensioners). Another tricky one is (the Australian Labor Party). To give that the full title it would be (The American Labor Party in Australia). If the Labor Party had its roots in Ephraim it would be The Australian Labour Party (with a 'u', just as it is in Britain). Some of the descendents; of the tribes-of-Israel are . . .

The tribe of **Manasseh** is basically the **United States**.

The tribe of **Ephraim** is mainly the English-speaking countries of the British Commonwealth. (This includes Canada, New Zealand, parts of South Africa and **Australia**).

The modern-day Jews (scattered among the nations) are known to be of the tribe of Judah, A large portion of them live in a place named Israel.

Other tribes are scattered among many nations. ~ Finland. ~ Holland (The Netherlands). ~ Norway and Iceland. ~ Ireland and portions of Denmark. ~ Sweden. ~ Switzerland ~ Belgium and Luxembourg.

Note; Denmark began as 'Dan's Mark'

Many gentile nations have mingled in these same countries. The tribe of Dan may have passed its 'use-by-date.' The other tribes are all **Israel** and all follow on from the promise.

So next time you are 'talking-down' to Israel; try to remember who you are talking about. The theme of 'twelve tribes' is consistent throughout The Book.

According to popular opinion; present day Judah are the only nation of Israel—overlooking the fact that the 'real Israel' includes *all twelve* **tribes**.

CHAPTER ELEVEN

'SANDY CLAWS'

As Richard has shown, future generations might learn to investigate the origin, of "ancient and mysterious tree-ceremonies," before they become "infected"

["Such free spirits should need only a little encouragement to break free from the vice of religion altogether]." [www.infoamerica.org/documentos_pdf/dawkins10.pdf]

Such an effort would end the process of passing on the **false**—from generation to generation. This chapter takes a look at a very popular character; (who hangs around shopping malls in our country). Some of this takes a look at where he originated from; and whether he belongs on your list of 'things-to-do.'

To make the task simpler; please take another look at the Instruction Book. I have already shown the "tree-ceremony" in chapter one. Now for further proof of deception; take a look at 'Revelation'; but first we need to be sure about who or what we are looking for . . . .

[Who are the "Nicolaitans"?]

[A mysterious group of wicked religious imposters; were the "Nicolaitans." What did they teach? Do they still exist today? Why should they concern you? In the warnings to the seven churches of

Revelation, we are told to beware of them. Why are they dangerous and how would you recognise them? . . .

In Revelation, chapter 2, we read of an enigmatic sect or group; called the Nicolaitans; who post a great threat to the churches of God. The Messiah, (Yeshua), says to the Ephesus church: "But this you have, that **you hate the deeds of the Nicoaitans, which I also hate**" (Rev.2:6).

[William F. Dankenbring:]

[www.triumphpro.com/nicolaitans.htm]

Notice that the church hated and rejected the doctrines {the practices}, of the Nicolaitans, {not the people themselves}.

Now that we have the background, we can easily work-out where the imposter comes from. It is of course "Old Saint Nicolas;" of the "Nicolaitans", also known as "**Old Saint Nick**." The name later developed toward [saint-nee-clause] and then [saintee-clause]. The most popular rendition of it (in our time) is [santa-clause]. We were all brought up with this 'idea.'

We do not need to pass on this distorted custom. If you would like to find out more about the **tree**; I suggest doing some research into ancient 'Asherim' objects. I cannot include the details of the Asherim or I will end up with an 'R' rating. It was the research of the tree-ceremony that led me to cease the 'practices of the Nicolaitans.' If that study isn't enough to stop you; then just think of how much better our economy would be; if we didn't have to 'over-spend' each year, and then go without; while trying to recover from 'the silly season'.

The second most popular festival (in the United States) is the Halloween nonsense. Australians now face the risk of catching this off-shore virus and passing it on; to our unfortunate 'generations.' American citizens spend over two billion dollars each year on this rediculous and annoying behaviour. Besides teaching children that it's alright to beg for something—rather than working to earn it—it is also turning them into extorsionists, demanding something-for-nothing; (or or I will play a nasty trick on you).

The Encyclopaedia Britannica says the following:

["Samhain (Celtic: 'End of Summer'), one of the most important and sinister calendar festivals of the Celtic year. At Samhain, held on November 1, the world of the gods was believed to be made visible to mankind, and the gods played many tricks on their mortal worshippers; it was a time fraught with danger, charged with fear, and full of "supernatural" episodes. Sacrifices and propitiations of every kind were thought to be vital, for without them the Celts believed they could not prevail over the perils of the season or counteract the activities of the deities. Samhain was an important precursor to Halloween."]

Hebrew: 2:14; . . . that through death he might destroy him that had the power of death, that is, the devil. (KJV)

The ancient Celts, (who actually believed they were worshipping the true God), were deceived into worshipping 'the god of this world'. The present ruler over the earth; is in fact "the father of lies and religious deception." The Celts also bring us the Celtic Cross which many "believers" hang on their walls, falsely thinking of it as a "religious" object. The thing which religious fundamentalists think of as a token of *their* religion; is in fact a reminder of human torment. The torment that was inflicted on an innocent victim; by a couple of thrill-seeking Roman soldiers. What if the instrument of death had been a baseball bat; (instead of a pole). Would the religionists then wear a miniature baseball bat—or hang such an image on their walls to commemorate the occasion?

CHAPTER TWELVE

'ISHTAH'

Thankfully; in Australia the Halloween bug is only at the worm stage. The second favourite (here) is the celebration for 'Ishtah.' This is by far the easiest **hoax** to investigate and expose. It obviously comes from ancient and myterious times. A time when 'the gods' had more than one name; (the name gradually changed over time—just like sandy claws). **Ishtah** (pronounced easter) is another name for **Ashteroth**.

The goddess Ishtar "the light-bringer", is the Babylonian high mother-goddess; the goddess of Fertility, Love and War.

1 Samuel: 7:3; And Samuel spake unto all the house of Israel, saying, If you do return unto the Lord with all your hearts, then **put away the strange gods** and Ashtaroth from among you, and prepare your hearts unto the Lord, and serve him only: and he will deliver you out of the hand of the Philistines. Then the children of Israel did **put away Baalim and Ashtaroth**, and served the Lord only. (KJV)

Another alias would be Astarte. According to scholar Mark S. Smith, Astarte may be the Iron Age incarnation of the Bronze Age. **Asherah**. Astarte is the name of a goddess as known from Northwestern Semitic regions, cognate in name, origin and functions with the goddess **Ishtar** in Mesopotamian texts].

[en.wikipedia.org/wiki/**Queen_of_heaven**(antiquity)]

(see also 'Ashera Pole' and 'Asherim')

As you can see from the above writings; the seemingly harmless celebration; known as easter; has a dark side to it. The **eggs** are obvious symbols of fertility, and we all known what **rabbits** represent. When you buy gifts for the 'queen of heaven'—you are in fact introducing a celebration of fertility. The only thing that hasn't been resolved; is whether your 'gifts' are for her 'fertility'; or for her 'not-fertility'. I am guessing it is the latter; because it makes sense for a 'queen of the night' to have it that way.

Now that I have put the 'Ishtah Eggs' into the correct basket; you might like to consider the remainder of the deception. The whole "Good Friday—Easter Sunday" tradition is a complete fabrication. The (churches) of this world rely on . . .

Confusion

Disagreement

Competition

The real text of the Instruction Book teaches us to recognise these 'conditions.' The Sun-day churches claim to be based on the truth; while actually thumbing their nose at the plain language; delivering a message to the reader.

The first message to consider is about the correct timing of events. Consider how many **hours** there are (from Friday afternoon to Sunday morning). By making a simple calculation; I arrive at (24+12) **thirty-six hours**. That is twelve for Friday night—twelve for Saturday (daylight hours)—and twelve more for Saturday night. The second message to consider is about the state-of-affairs which is clearly portrayed early; (daybreak); on Sunday Morning. Once the two items are resolved; we can 'do-the-math.'

The note above gives us the apparent 'official' version; which is obviously a time period of thirty six hours. Please take a careful look at the real story and do your own calculations—to find a period of almost exactly **seventy-two hours.**

Jonah 1:17; . . . And Jonah was in the belly of the fish three days and three nights.

Matthew 12:38; Then certain of the scribes and of the Pharisees answered, saying, Master, we would see a sign from thee. But He answered and said unto them, an evil and adulterous generation seeketh after a sign; and there shall no sign be given to it, but the sign of the prophet Jonas.

Matthew 12:40; For as Jonas was three days and three nights in the whale's belly; so shall the Son of man be **three days and three nights in the heart of the earth**.

The above passage from Matthew cannot be weasled away. Three days and three nights is always going to add up to **seventy-two hours**. So what led to the Friday tradition? It 'arrived' because of the refernce to a sabbath-day; (the following day). The confusion (for some) came from the fact that there was an extra 'day off' during the week. More confusion was added when Bible translators (possibly under pressure) changed the word Passover to [easter]; thus disregarding the meaning of the written Hebrew text.

To help keep track of the days we need to look-back to the prophet Daniel, who lived in the period of the Babylonian captivity. The writing of Daniel was in the period (605-538 BC)

Daniel: 9:27; And He shall confirm the covenant with many; for one week: and **in the midst of the week** he shall cause the sacrifice and the oblation to cease. So it is clear that the "old covenant" sacrifices would be (done-away-with) and the one sacrifice would take their place, and we also see that it was to occur in the middle of the week.

The crucifixion occurred on Passover day—the 14th of Abib or (Nisan), (the first month in the Sacred Calendar). This event occurred on a Wednesday in the year 31 AD! According to the Roman calendar; it was Wednesday the 25th April. The time was announced as between the ninth and twelfth hours—(which refers to daylight hours). Thus the time of death is just before sundown. The following day was a part of an annual festival (it was a day-off and the shops would be closed). The Thursday Sabbath for Passover is the thing that caused the confusion (for those in the past who couln't read or didn't care). Now counting forward three days and three nights (from Wednesday afternoon) we reach Saturday (The Sabbath day) and the time of ressurection is just before sundown.

Now for the state of affairs on Sunday at sunrise.

Mark: 16:1; [And when the sabbath was past, Mary Magdalene, and Mary (the mother of James), and Salome, had bought sweet spices, that they might come and anoint him. And **very early in the morning the first day of the week**, they came unto the sepulchre at the rising of the sun. And they said among themselves, Who shall roll us away the stone from the door of the sepulchre? And when they looked, they saw that the stone was rolled away: for it was very great. And entering into the sepulchre, they saw a young man sitting on the right side, clothed in a long white garment; and they were affrighted. And he saith unto them, Be not affrighted: Ye seek Jesus of Nazareth, which was crucified: **he is risen; he is not here:** behold the place where they laid him] (KJV).

So for (any) reader, the message is quite plain—by daylight on the first day of the week, it was already over-and-done-with. Those who attend a sunrise service on easter Sunday are running on a completely false idea.

Alexander Hislop elaborated on the origin of easter (or astarte) in his book The Two Babylons: "It bears its Chaldean origin on its very forehead. Easter is nothing else than astarte, one of the titles of beltis, the queen of heaven . . . That name, as found by Layard on the Assyrian monuments, is Ishtar. In nearly all Semitic dialects, "ishtar" is pronounced "easter." Easter festivities extensively refer to celebrating the personage Ishtar, Ashtoreth, or the "queen of heaven," who has many interchangeable (false) names. Each year citizens in pagan nations celebrate her son's death and return during spring.

What does this message sound like (to you)?—Spring is in the air! Flowers and bunnies decorate the home. The children paint beautiful designs on eggs dyed in various colors. These eggs, which will later be hidden and searched for, are placed into lovely, seasonal baskets. The wonderful aroma of the hot cross buns waft through the house. The whole family picks out their Sunday best to wear to the next morning's sunrise worship to celebrate the resurrection and the renewal of life. Everyone looks forward to a succulent ham with all the trimmings. It will be a thrilling day. After all, it is one of the most important religious holidays of the year.

Sounds like 'Easter', right?—This is actually a description of an ancient Babylonian family—2000 BC—honoring the resurrection of *their* god, Tammuz, who was brought back from the underworld by his mother/wife, Ishtar. Throughout the centuries, millions of people have been persuaded to believe that easter's purpose is an honourable one; Yet this age-old global tradition can be traced back thousands of years (before its 'supposed' beginning).

For the first two hundred years of European life in North America, only a few states, mostly in the South, paid any attention to Easter. After the Civil War; Americans began celebrating this holiday: "Easter first became an American tradition in the 1870s The original thirteen colonies (of America) began as a 'christian' nation, they did **not** observe easter within an entire century of its founding.

Perhaps early American citizens knew of the promises made to Abraham; perhaps they were avoiding easter for a good reason!

CHAPTER THIRTEEN

THE PATHS OF THE SEA

This is a true story of (recent) scientific discovery. There are relevant things to consider, while reading this section. The Book known as 'The Book of Psalms;' was written by various authors; King David produced nearly half the contributions. The text quoted below was produced in the time of Moses (**1400 BC**). Science discovered the 'fact' in recent times. Also, if this text really was produced; only to suit The Church (and in more recent times); then the person doing so may have risked being wiped out, just for thinking something so blatantly scientific, (and so useful).

When 'Matthew F. Maury' was confined to bed; (early 1800's); he asked his young daughter to read to him. She opened up at Psalms 8: and verse 8: and started reading; "'the fowl of the air, and the fish of the sea, and whatsoever passeth through the **paths of the sea**" (KJV). (See note by K. STILES; shown below). Later, Matthew was nicknamed "Pathfinder of the Seas" and "Father of Modern Oceanography and Naval Meteorology" and later still, "Scientist of the Seas," due to the publication of his extensive works in his books, especially *Physical Geography of the Sea* (1855), the first extensive and comprehensive book on oceanography to be published. Maury made many important new contributions, to charting **winds** and **ocean currents**, including ocean lanes, for passing ships. The following paragraph shows a little, about the 'style' of Maury's work.

[This idea of divine order and design occurs again and again in the book like the motive in a piece of music; In fact, Maury, was well read in the Bible, (quotations from which appear in his writings by the dozen). He had very definite ideas about the relation between science and the Bible, and declared that it was his rule never to forget who was the Author of the great volume which Nature spreads out before men, and always to remember that the same being was the author of the Book which revelation holds forth for contemplation. It was his opinion that, though the works were entirely different, their records were equally true, and that when they bear upon the same point, as they occasionally do, it would be impossible for them to contradict each other. **If the two cannot be reconciled, the fault therefore is in man's weakness and blindness in interpreting them aright**].

[www.archive.org/stream/matthewfontainelewi/matthewfontain emlewi_djvu.txt]

["A Brief Sketch" of the Work of Matthew Fontaine Maury; By Richard Launcelot Maury 1915 AD]

The Introduction to the above:—

WHEN I took charge of the Georgia Room, in the Confederate Museum, in Richmond, Virginia in 1897, I found among the De Renne collection an engraving of the pleasant, intellectual face of Commodore Matthew Fontaine Maury, so I went to his son, Colonel Richard L. Maury, who had been with his father in all his work here, and urged him to write the history of it, while memory, papers and books could be referred to; this carefully written, accurate paper was the result.

At one time, when Commodore Maury was very sick, he asked one of his daughters to get the Bible and read to him. She chose Psalm 8, the eighth verse of which speaks of "whatsoever walketh through the paths of the sea," he repeated "the paths of the sea, the paths of the sea, if God says the paths of the sea, they are there, **and if I ever get out of this bed I will find them.**"

He did begin his deep sea soundings as soon as he was strong enough, and found that two ridges extended from the New York coast to England, so he made charts for ships to sail over one path to England and return over the other.]

[The proceeds from the sale of this little pamphlet will be used as the beginning of a fund for the erection of a monument to Commodore Maury in Richmond].

[KATHERINE C. STILES.]

CHAPTER FOURTEEN

THE CYCLE ENDS

The 16th century Italian philosopher (and former Catholic priest) Giordano Bruno was charged and then killed—for a stubborn adherence to his (then) unorthodox beliefs—including the ideas of the universe as infinite space, and the idea that 'other solar systems' exist. Art historian Ingrid Rowland vividly recounts Bruno's journey through a quickly changing Reformation-era Europe, where he managed to stir up controversy at every turn. Having a habit of calling schoolmasters "asses," Bruno was jailed in Geneva for slandering his professor after publishing a broadsheet, listing 20 mistakes the man had made in a single lecture.

Bruno's adventures in free thought ended when The Roman Inquisition declared him "an impenitent, pertinacious, and obstinate heretic," to which he characteristically replied, "You may be more afraid to bring that sentence against me than I am to accept it." In 1600 the inquisitors sentenced him to death. (At least his universe survived).

[Ingrid D Rowland: "Giordano Bruno—Philosopher and Heretic"]

Besides infinty; the other thing that was feared by superstitious people; was the number zero. It just didn't seem right to need a number that represented nothing. This lack of zero held-back the nations concerned for many years. (We certainly had no hope of a decimal system; no hope of modular arithmetic; and therefore no hope of electronic calculations—without the understanding of the value "zero." Also; on the subject of superstition—Had 'The Book of Psalms' been written (by anyone at all) in the years; say 100 to 1500 AD; the Author and the book would have been in deep trouble for the science-like thoughts it included.

Job: 26:7; He stretcheth out the north over the empty place, **and hangeth the earth upon nothing**. This idea of hanging something could potentially lead to wanting "corners" instead of "quarters (see ". . . winds")); (perhaps corners would make it easier to hang)?

Job may be one of the oldest Books . . . (and it refers back; to an even earlier time); probably witten by around 2000 BC. Please keep in mind that Albert Einstien first wrote about **gravity** approx. 3900 years later.

Exodus 34:21; six days thou shalt work, but on the seventh day thou shalt rest: in earing time and in harvest thou shalt rest. And thou shalt observe the feast of weeks, of the firstfruits of wheat harvest, and the feast of ingathering at the **year's end**. The words year's-end are translated from the Hebrew: (shaneh) {shaw-neh} in plura or (feminine) (shanah) {shaw-naw'}; **a year** (as a revolution of time); followed by the Hebrew word: (tquwphah) {tek-oo-faw'} or (tquphah) {tek-oo-faw'}; **a revolution**, i.e. **(of the sun-course)**, (of time) lapse: **circuit**, come about, end. Please note: The ancient (Hebrew) dictionary clearly links a year with a revolution around a 'course' (which takes one year to complete). And it had this definition thousands of years before the 'science' was published. There certainly was no reason—even in ancient times—to believe that the sun travelled across the sky.

Ecclesiastes: 1:6; (The wind goeth toward the south, and turneth about unto the north; it whirleth about continually, and the wind returneth again according to His circuits). The Hebrew for 'circuits:' (cabiyb) {saw-beeb'} or (feminine) cbiybah {seb-ee-baw'}; (as noun) a circle, neighbour, or environs; but chiefly (as adverb, with or without preposition) around: (place, round) about, circuit, compass, on every side. When combined with; Revelation of John 7:1; And after these things I saw four angels standing on the four [corners] of the earth, holding **the four winds of the earth**, that the wind should not blow, on the earth nor on the sea, nor on any tree. Note; the key word here (and perhaps mistranslated); Hebrew: (gonia) {go-nee'-ah}; an angle, corner or **quarter**. So today; one might accurately translate it to read; **on the four 'quarters' of the earth**, holding back the four winds. The actual message was recorded thousands of years ago, at a time when humans had little knowledge of the weather (no high-flying aircraft) and (no sophisticated test-equipment)—yet clearly in agreement with something recently discovered by science, as follows . . .

[A narrow current of strong wind circling the Earth from west to east at altitudes of about 11 to 13 km above sea level. There are usually **four distinct jet streams**, two each in the Northern and Southern hemispheres. Jet-stream wind-speeds average 56 in the summer and 120 km/hr in the winter. They are caused by significant differences in the temperatures of adjacent air masses. These differences occur where cold, polar air meets warmer, equatorial air, especially in the latitudes of the westerlies].

[See page > www.thefreedictionary.com/jet+stream]

For those interested, I have included some (very brief) history of churchy customs. In the year AD 324 the Roman Emporer Constantine— (presumably after seeing a vision)—established "churchianity" as the official religion of the realm. A year later The Council of Nicea was established (to govern disagreements). In the year AD 363 The Council of Laodicea declared Sunday to be the official day for christian worship. (In place of the Bible's 'seventh day' Rest). During this period of earth's history it was close-to-impossible for anyone to read and obey The Instruction Book; persons keeping The Sabbath faced 'trial' with a certain death-sentaence. (As you can imagine); the influence of the Roman Church grew over the years. By AD 476 the wheels began to fall off. They were eventually defeated by barbaric tribes. This event rocked Europe (and the Vatican). After this defeat, the papacy continued (out of sight and against the invaders). By AD 554 Justinian restored an empire; known as "The Holy Roman Empire." This man-made existence was then named and promoted as *God's kingdom on earth*. Note; 1 Corinthians 15:50; Now this I say, brethren, that *flesh and blood cannot inherit the kingdom of God* . . . (KJV)

With the now strong ties between Church and State; the 'State' carried out the wishes of the extremists (inside the establishment); and in turn the Church gave the ruler of the day, some level of credibility with the population. So at this point in time; The Vatican became the new kid on the block. Because of this power; (Church and State); the Empire remained strong for another two hundred years.

So, all those centuries ago, we see the Pope of the day; (along with some good folks to do the killing); not only gave us Sunday as (our) new rest day; but at the same time gave us the slaughter of anyone who dared to keep the commanded "Seventh Day."

Ten questions that "came to mind" (while watching).

Does the book support "The tree-ceremony"—NO!
Does the book support "Praying to Mary"—NO!
Does the book support "Floating to heaven"—NO!
Does the book support "Bad-time in hell"—NO!
Does the book support "Halloween"—NO!
Does the book support "Easter"—NO!
Does the book support "Good-Friday"—NO!
Does the book support "Sunday services"—NO!
Does the book support "a 'triune' or 'trinity'"—NO!
Does the book support "a young-earth"—NO!

The relationship, between the Roman Church and the other 'Sun-Day' churches, can be compared to a Mother; the "inventor" [of the new UNIVERSAL church] and her "disobedient daughters." The daughter churches were told to "stay away from the Sunday;" to go back to the old-fashioned "God of the Bible." The amount of [other peoples] blood sweat and tears that went into making Sunday; meant the Catholics wanted Universal rights to Sunday—so that all Sundaykeepers would be Catholics. Merely by chance; the Catholic writers were in fact giving 'good advice;' to the other Sunday followers; (the ones they 'thought' they were shunning). What a terrible dilema for the remainder of churchianity. If they stay on Sunday; they are treading on Universal toes. But if they turn around and run back to the original text; they would be all–but admitting that Sunday worship is a hoax; therefore leading to the ire of the Catholic writers. For them it's a lose-lose situation; with no backing down.

THE BOOK

The question of the Authenticity of 'The Book.' Do we have a reasonable 'copy' today; showing the original 'intent' and the original 'meaning?' This section takes a brief look at what has happened to the various books; along the way. With these accounts; you'll find some ideas about what went right—and some things that were changed. Many millions today believe that the Catholic Church "canonized" (approved) the Bible for our use. I hope to show a different case; (or at least encourage you to investigate further). The "pre-catholic" books were preserved reasonably intact. Only the 'order' of the books was 're-arranged.' By the time 'The King James Version' was prepared; the content was still (surprisingly) as accurate as possible (considering translation issues and the un-authorised additions). For further interest; take a look at . . . [en.wikipedia.org/wiki/Dead_sea_scrolls]

Today's students will surely debate the official "age" of the Dead Sea Scrolls. The (200 BC) dating of these was determined by the type of parchment and by the Square Hebrew characters used for writing. Also note that some of the [paper] was thought to be from hundreds of years earlier. It is important to note that the scrolls were being **copied**. The workers had access to a store of preserved [paper] and would chose according to what was being copied. The scrolls may have been hidden away (in the caves) any time from 200 BC through to 70 AD (and still appear to be from an earlier time). Note also that after 70 AD; the Essenes and The Sadducees were practically non-existent. Another cause for debate will be the carbon dating of some of the [pages] (which came out at 2000 BC). The Book of Job was originally written around 2000 BC. Thus; an original [page] from the (book) of Job would be expected to show a much earlier time-scale.

WE all know that 'Wiki' articles can be corrected or even removed at some future date; but there it is—for today's generation (i.e. preserved by "Wiki" for us to read). From it; I see the Romans out to destroy the work. Also; note that the Essene Community itself (the ancient community that was being destroyed) was a community of Jews trying to preserve The Scrolls. Note also that (some accounts) have the Essennes preserving the originals as; "backup copies;" but also producing some counterfiet [pages] to reinforce some of their own ideas. The Essennes had rules and laws governing their everyday lives in a similar fashion to the other groups; such extra restrictions were (ok) for them; (it allowed them to perform a task); however such restrictions were not to be 'peddled' as an idea for everyone to obey. Do we encounter "some accounts," in history? A recent example of (some accounts) occurred (one single event); when North Korea decided to launch an experimental 'carrier' craft. One record of history proudly declares the carrier craft to be an outstanding success. However; there were "some accounts" of a dismal failure. Readers (of the future) might 'see' the success story. Then be able to dig a little further and find 'some accounts' which show a spectacular failure. Just a little aside here; whenever a 'genuine' copy was made; they would carefully copy the text (writing with [pen] and ink on specially preserved [paper]; then follow each [page] with a manual word count; and letter count). So we can see that some 'scribes' who were set-apart for this task intended to work with great accuracy.

There were three main groups of religionists in Judah; (particularly in the period from 200 BC to 70 AD); these were The Pharisees (the ones who 'thought' they were somebody) then there were the Sadducees (the Elite), and the Essennes. You see the two main groups mentioned often. You also see "The Scribes and Pharisees;" shown together; where they have something in common. (i.e. They were both adding burdensome rules to daily life; without good reason). These

'Scribes' were definitely around at the same time—and may have been the 'Essenes.' These groups would have been well defined and well known (at the time). An interesting point arises here. The Pharisees of the time; who enforced their own rule of law (i.e. a 'burdensome' interpretaion of their own 'man-made' law); were named as 'Separatists.' The majority of the population (Judah) were none of these groups. But they certainly needed to obey certain rules and laws whenever the "authorities" were around. The Samaritans were also in the business of producing counterfiet 'Books.' It was Ezra who brought about the change to block Hebrew for genuine texts to make a clear distinction between one 'copy' and another. Simon Magus carried on the deception (from where the Samaritans left off).

Taking a look at the Old Testament we can discover that the 'content' (of the books) is fairly accurate. First, take a look at the definition of the word "Tanakh."

[Ta·nakh or Ta·nach (tä-nä); noun. The sacred book of Judaism, consisting of the Torah, the Prophets, and the Writings; (the Hebrew Scriptures). [Acronym from the initial letters of the Hebrew names for the Torah, the Prophets, and the Writings; t(ôrâ), n(dî'îm), k(tûbôt).]

[The American Heritage® Dictionary of the English Language, Fourth Edition copyright ©2000 by Houghton Mifflin Company. Updated in 2009. Published by Houghton Mifflin Company. All rights reserved].

From this we start to look at the 'arrangement.' We find that The order of the books was already upset by the time of the King James Version. For example one of the original "Books" carried the title "The Twelve." Or "The Twelve Prophets." (One book enclosed twelve smaller books), (sorted by a single subject). Many [scholars] publicly acknowledged that there were 22 books in the Hebrew Scriptures: Origen (AD 210), Athanasius (365), Cyril of Jerusalem (386), and Jerome (410).

Here then; is the correct order for the Old Testament Books. (T); We first list the books of the law, also known as the Torah or Pentateuch.

The Law of Moses (five books): containing;

Genesis; Exodus; Leviticus; Numbers and Deuteronomy (Note that the first section, has not changed). The changes creep in for the second and third sections.

Now, the original order of the Prophets. (Note how some books have been divided and then sub-divided:
The Former Prophets (2 books):

Joshua and Judges (combined into one)
I-II Samuel and I-II Kings (all four combined into one)

The Latter Prophets (four books):

Three major prophets: Isaiah, Jeremiah and Ezekiel

Plus "The Twelve" (Prophets); one book; (consisting of twelve prophetic books combined into one)

Hosea, Joel, Amos, Obadiah, Jonah, Micah, Nahum, Habakkuk, Zephaniah, Haggai, Zechariah and Malachi.

The third division; known as the Psalms (the Writings)
The Former Poetic Books (three books):

Psalms, Proverbs and Job

The Festival Books (five books):

Song of Solomon, Ruth, Lamentations, Ecclesiastes and Esther

The Latter Restoration (three books):

Daniel (one)
Ezra-Nehemiah (combined into one)
I-II Chronicles (combined into one)
(Note; the original order is chronological).

Keep in mind it was mainly the order that was "mucked up;" the actual contents remained fairly intact; preserved by Judah in the Hebrew language; (as shown elsewhere).

Now for the order of the New Testament Books.
(containing 27 books in four sections.)

Gospels (and Acts):   Matthew, Mark, Luke, John and Acts
General Epistles: James, I-II Peter, I-III John and Jude.

They were intended for the general Church of God and not addressed to any specific congregation. They largely contain general information.

Paul's Letters to Specific Churches:

Romans, I-II Corinthians, Galatians, Ephesians, Philippians, Colossians and I-II Thessalonians.

Paul's General Letter: Hebrews

Paul's Pastoral Letters: I-II Timothy, Titus and Philemon

Other Writings of John: Revelation.
Revelation is from the Greek: (apokalupsis); {ap-ok-al'-oop-sis} Meaning: disclosure: [appearing, coming, lighten, manifestation, be revealed, revelation.] Revelation is associated with two names—"The

Apocalypse," which is correct as shown; and sometimes; "Armageddon," which is false. (Based on the use of the word "Megiddo"). "The Battle of Megiddo" actually took place in the 15th century BC!

Often known as the "prison epistles," Paul wrote the following books while in the slammer:

(Ephesians, Philippians, Colossians and Philemon.)

Now adding up; we find 22 books of the Old Testament and 27 books of the New Testament, a total of 49 books—representing absolute completion. Out of envy, the Jews (of the second century) altered the number of "their" books to 24 (by splitting two books)—to erase this significance.

Even in the face of 'many changes,' the Scriptures remain largely 'intact,' though the order of the Old Testament has been rearranged; (primarily by the Roman Catholic Church, and following the order of the 'corrupt' Septuagint version). They also rearranged the New Testament, to exalt Rome. But again, this does not mean they decided or established the contents! So; if the "non-catholic" book has been preserved for this generation, why are there differing translations and which is which? We simply need more than one, because some 'loss of clarity' comes from world-view translations, such as 'vultures' in place of 'eagles'. Some translations are done word-for-word; some translations are from meaning to meaning; and then put in context (in English). Any method can cause problems if the scholars are language experts with pre-conceived ideas. Looking-back to the Greek, (from my own point of view), is not always as successful as it could be. I am working backwards; from English to the Greek. A much greater step back; (towards the original intent); would be to acquire an earlier text, (assuming I could read the Greek) The safest and most honest way to read; is to consider the context of any particular section. Sometimes the King James Version will have a word (or two) that could be unsound (e.g. corners) then another translation might make it clear. Many times, the reverse is the case (e.g. vultures). The context; (of the vultures); is discovered, by knowing two things. The 'body,' is not a body that has died—or is about to die, it is in fact a body of many people gathered together. The bones being gathered by the eagles; are dead bones (with no food-value for vultures to be concerned with). As for the corners—the same word is used elsewhere—and is translated 'quarters.' As in the "Revelation of John"; chapter 20; ". . . And shall go out to deceive the nations which are in the **four quarters** of the earth" (KJV).

As I have said, you do not need to be an expert in translation to be able to read and understand; you only need to 'desire' the true picture; and it is there in plain sight. For every counterfeit out-there, there can be found an opposing explanation; (which is generally, one that makes a lot more sense).

Sometimes a simple comma can cause confusion. Also; the scrolls did not have chapter and verse divisions. (Division into chapters could break-up a story in an unwanted way). In the following; a simple comma changes the entire sentence . . . The comma, which follows a lead-in statement, "Verily, I say unto you . . ." was added and misplaced. It changed His entire meaning. The original Greek, (the language of the New Testament), did not use certain punctuation, such as commas and quotation marks. Translators (using their own discretion) added them later. The correct rendering is, "I say unto you today [in other words, "I'm saying this now"], you shall be with me in Paradise." When rightly concerned about the context it's easy to see a future event; (which still hasn't occurred to this day); rather than an entirely impossible event. We know for sure that the thief wasn't going to float-on-up-to-heaven. And; since Christ did not condemn the thief, he simply assured him of a future event. He could not have literally meant the same day. He had obviously not yet died; i.e. The counting of "three days and three nights" had not yet started; and the earth was definitely not a 'paradise.'

The text found at **1 John 5:7 and 8**; is a lovely churchy sounding piece that trinity-believers love to read. However the lovely churchy text **is as fake as a two-bob watch.** Actually added in 800 AD to directly support a supposed 'trinity.' Transcribers who believed in the trinity could find no scriptural support—so they added some words to support their own bizarre beliefs. The inserted text (the last half of verse seven and all of verse eight); is pure fantasy! Those who use these verses to support the 'trinity;' are either unaware that the passage was altered, or they are aware but feel that their use serves a 'greater good.'

Many Bible margins directly state the truth of the passage. For example, the New King James Version (margin) reads like this: "NU, M [versions] **omit the rest of v. 7 [after "record"] and through to the end of v. 8**, a passage found in Greek in only four or five very late mss. [manuscripts]."

The print that refuses to fade away:—In an earlier quote I mentioned the ten rules (commandments). Now take a look at what is recorded and then carefully compare it to the view held by the present 'churches'. But first let The Book show if such changes are truly lawful [or just plain awful].

Matthew 5:18; For verily I say unto you, Till heaven and earth pass, **one jot or one tittle shall in no wise pass from the law**, till all be fulfilled (KJV).

Psalms 89:34; My covenant will I not break, nor alter the thing that is gone out of my lips (KJV).

Daniel 7:25; . . . he shall speak great words **against** the most High, and shall 'wear-out' the saints of the most High, **and think to change times and laws** (KJV)

1 John 5:3; For this is the love of God, **that we keep His commandments**: and His commandments are not grievous.

In the following section there are some 'pairs.' The first of each, is the true record from The Book; the second; [enclosed]; is the twisted and distorted version—(which is often peddled as the truth).

One: Thou shalt have no other gods before me.
Changed to . . .
[I am the Lord thy God: thou shalt not have strange gods before me].

Two: Thou shalt not make unto thee any graven image, or any likeness of any thing that is in heaven above, or that is in the earth beneath, or that is in the water under the earth: Thou shalt not bow down thyself to them, nor serve them: for I the Lord thy God am a jealous God, visiting the iniquity of the fathers upon the children unto the third and fourth generation of them that hate me; And shewing mercy unto thousands of them that love me, **and keep my commandments.**

Changed to . . .

[Thou shalt not take the name of the Lord thy God in vain].

Three: Thou shalt not take the name of the Lord thy God in vain; for the Lord will not hold him guiltless that taketh his name in vain.

Changed to . . .

[Remember that thou keep holy the sabbath day].

Four: Remember the Sabbath Day, to keep it holy. Six days shalt thou labour, and do all thy work: But the seventh day is the sabbath of the Lord thy God: in it thou shalt not do any work, thou, nor thy son, nor thy daughter, thy manservant, nor thy maidservant, nor thy cattle, nor thy stranger that is within thy gates: For in six days the Lord (made) heaven and earth, the sea, and all that in them is, and rested the seventh day: wherefore the Lord blessed the sabbath day, and hallowed it.

Changed to . . .

[Honour thy father and thy mother].

Five: Honour thy father and thy mother: **that thy days may be long** upon the land which the Lord thy God giveth thee.

Changed to . . .

[Thou shalt not kill].

Six: Thou shalt not kill.

Changed to . . .

[Thou shalt not commit adultery].

Seven: Thou shalt not commit adultery.

Changed to . . .

[Thou shalt not steal]

Eight: Thou shalt not steal.

Changed to . . .

[Thou shalt not bear false witness aginst they neighbour].

Nine: Thou shalt not bear false witness against thy neighbour.

Changed to . . .

[Thou shalt not covet thy neighbour's wife].

Ten: Thou shalt not covet thy neighbour's house, thou shalt not covet thy neighbour's wife, nor his manservant, nor his maidservant, nor his ox, nor his ass, nor any thing that is thy neighbour's.

Changed to . . .

[Thou shalt not covet thy neighbour's goods].

The first of each is from the Instruction Book and will never change. The second; [below each one] is taken from the . . . [Convert's Catechism of Catholic doctrine. P 37; published by B. Herder Book Co. (1921)]

Boot 'em Out!

The well known presidential historian Peggy Noonan summarised mankind's complex, jumbled history in this way: "In the long ribbon of history, life has been one long stained and tangled mess, full of famine, horror, war and disease. We must have thought we had it better because man had improved. But man doesn't really 'improve,' does he? Man is man; Human nature is human nature; the impulse to destroy co-exists with the desire to build and create and make better." ["America's Age of Uncertainty;" Knight-Ridder; 2001.]

People on either side of religion are confused about their own condition. Those who claim to be religionists have no respect for the Book, yet they claim to learn from it. To the religionists I say . . . "either read the Book and make the changes, or tell your customers the truth about the way you operate."

Those who proudly claim to be athiests are only part-time athiests; they firmly hold-on to their annual festivals and rituals without taking a second thought. Such rituals, when carried out each year, arc in fact 'their' religion. To the athiest I say . . . "either give up your pagan rituals or admit that you prefer to remain steeped in religion." A person from either of these groups would think I am strange, simply because I refuse to participate in their annual 'holy'-days.

The future will decide if our nation manages to eliminate these ancient; mysterious customs. It would only take a generation or so to achieve a good outcome. Future Australians might consider providing (and supporting) the best available education for students. Do you want Australian students being taught absurd; (man-made); ideas that will have to be 'shaken off' as soon as possible. Let's get rid of these 'Wallys' from our world and make way for some straight talking and clear thinking.

Whether we like it or not, we learn and remember things without even trying. For example; I have never chosen to watch "Home and Away" or "The Bold and The Beautiful," yet somehow I know that the town in 'Home and Away' is named 'Summer Bay,' and I know that a person named 'Ridge' is one of the characters in the 'The Bold.' When pressed I know that these (facts) are from a fictional source, but such is not a priority. I manage to recall the above 'names' as easily as the name of the American president. The point is, my brain is not sorting the information by the 'reality' field. When students 'pick-up' the rubbish that is being peddled around the world; under the guise of religion; they tend to remember—and then pass it on as 'information.' The cycle of lies and deception goes on—unchecked. Whenever you hear some 'churchy' sounding words from someone, try to keep this in mind, try to see through the garbage, try to discover the truth; you've got nothing to lose—except a tiny chunk of your previous world-view. You can someday teach your children the truth about the way religion operates, to help them avoid the serious problem of passing on the virus. You can someday play your part in booting-out these false teachers from our classrooms—for good! Doing these things will seem hard at first, but after a few years, the pagan rituals come and go without a second thought.